Copyright © 2016 by sbBooks

All rights reserved. No part of this publication may be reproduced, distributed, or transmitted in any form or by any means, including photocopying, recording, or other electronic or mechanical methods, without the prior written permission of the publisher, except in the case of brief quotations embodied in critical reviews and certain other noncommercial uses permitted by copyright law. For permission requests, write to the publisher, addressed "Attention: Permissions Coordinator," at the address below.

Bill the Geek
bill@billthegeek.com

Table of Contents

Copyright © 2016 by sbBooks..2
Introduction..7
Basics...8
   Equipment needed..8
   Equipment for brewing:..9
      Heat Source:...9
      Boil Pot:..10
      Mesh strainer:..11
      Kitchen scale:..11
      Digital timer:..12
      Thermometer:..12
      Long-handled spoon:...13
   Equipment for fermenting:..14
      Airlock:...14
      Bung:..14
      Fermenter:...15
      Hydrometer:..16
      13-inch Auto-siphon:...17
   Equipment for bottling:...17
      Bottling Bucket:...17
      Bottle filler:..17
      Bottles:...18
      Bottle caps:..19
      Bottle capper:..20
   Cleaning equipment:...20
      Classic cleaner:..20
      Sanitizer:..21
      Bottle brush:..22
   Additional equipment:...22
      Measuring cup:..22
      Strainer:...23
All about ingredients..23
   Malt:..24
   Hops:..29

"Humulus Lupulus"!..29
Leaf vs. Pellet hops...31
NOTE: When to add hops?...32
Yeast:..33
Water:...35
A Quick Peek..37
Timeline to brewing beer...37
The Brew Day:..38
The timeline of a Brew Day:...39
The first hour; make the mash..39
Second hour; the sparge..40
Third hour; the hop boil..40
Fourth hour; the cooling..41
4 ½ hours – the transfer..41
The Brew Day..42
Brew, Walk and Talk like a genuine Brewer...................................42
Lesson one – all about mash (only if using grains).........................43
Brewer's special terms:...43
Action and reaction: what happens?...44
Mash issues..45
Lesson two – all about sparge (only if using grains)......................47
Brewer's special terms:...47
Action and reaction: what happens?...48
Proper sparging:..49
Lesson three – making wort from a malt extract.............................52
Brewer's special terms:...53
Action and reaction: what happens?...53

Lesson four – all about hop boil..55
Brewer's special terms:...55
Action and reaction: what happens?...56
Lesson five – pitch the yeast..59
Brewer's special terms:...59
What happens to your beer now?...64
Primary fermentation...64
First 12-24 hours...65

Next 1-3 days..65
   Transferring beer from the primary to the secondary.....................68
   Is transferring really necessary?..68
   Siphoning beer..70
   Secondary fermentation..73
Bottling..75
   What happens to your beer while in the bottle?...............................77
Beer Recipes..78
   Classic Lager..78
   Robust beer..80
      Ingredients:..80
      Directions:...80
   American IPA...82
   Belgian beer...85
   Red ale...87
   English ale..90
   California lager...93
What to serve with your Beer?...97
   Hot Chicken wings...97
   Scotch Eggs...99
   Spiced Pork belly...101
   Grilled Lamb chops...103

Introduction

So, you have decided to make your own beer? Aren't you clever, you little weasel? Yes, we know you need to have a lot of beer for your friends, family, and you of course, because you love beer, otherwise you wouldn't be reading this book.

As you know, milk is for babies and when you grow up you drink beer. As some say beer is a proof God loves us and wants to make us happy.

Beer brewing is not very easy, but not too difficult either. Beer brewing is very satisfying with only a few critical issues. As long as you follow the beer brewing steps, we are sure the success is guaranteed.

Beer brewing is not limited only to only one type of the beer. As you continue to make your beer, you will feel more confident and be able to experiment with different flavors.

You will notice we have made two types of beer brewing for the beginners in this book:

> •One is for those who do not have enough space, or time and is with malt extract. Second is for those who want to make beer from a scratch – with real grains.

Whatever you choose it will not influence your beer in a wrong way. They are just longer or shorter version, but the beer gained with any method will be simply amazing.

With the first brew, you will see there is nothing too difficult in a beer brewing. Make sure you follow all the suggestions in this guide. If you have any doubts, at the end of each chapter we have described potential issue, and a solution for many.

Before you start, make sure you have all the equipment, all the ingredients, and time needed to pull this off.

We believe you are equipped with all the information, and that this guide will give answers to all your questions.

Happy Beer Brewing!

Basics

Equipment needed

Equipment is mandatory! You will not be able to make anything nevertheless the beer without equipment. So, what does a beer equipment include? You may see the images of different canisters, pots, and so on the web, but believe us, that sort of equipment is not good for the beer making unless you want to make some crappy beer. Just to know, you will not be able to brag yourself with a crappy beer, so investing in a quality equipment is very important. We cannot emphasize how important it is, so we will say it this way, IT IS VERY IMPORTANT!

Equipment for brewing:

Heat Source:

It all starts with a fire; a heat source is the first thing you need to have. And we are not talking about plain camp fire, but a reliable heat source, a type of heat that can bring at least 3 gallons of water to a boil. Stove in your kitchen will usually do the trick, but if you are preparing your beer in a comfort of your garage, then you need to have a reliable heat source.

Boil Pot:

You must have quality boil pot. The boiling pot is a place where all the magic happens. The entire boiling process with all ingredients is happening in the boiling pot. As a beginner, you will probably start with smaller quantities, but as your appetite grows (your family and friends appetite too) the higher amount of beer will have to be made. Therefore, investing in a quality and durable boil pot is imperative.

Mesh strainer:

You need a mesh strainer or fine colander so you can separate grains for the liquid you have created during the brewing process. The ideal size is the one that can hold up few pounds of wet grains. The mesh strainer should be able to fits easily inside the oil pot and your fermentation bucket.

Kitchen scale:

Precision is important in any job, especially beer brewing. You can choose a classic kitchen scale, but we suggest that you get an electric one, just to increase precision. The scale does not have to be all shiny and fancy, but the one that will do the job and accurately measure down to 1 gram.

Digital timer:

We are only mentioning this item, but anything can be a timer. If you have an Android or IOS, then you already have a built-in timer. Again, this does not have to be fancy, just easy and practical to use, and of course to beep when the time is out.

Thermometer:

A reliable instant-read thermometer is a must-have piece of brewing equipment. For an accurate brewing digital thermometer is the best and investing in quality one is a smart thing to do. You can find various thermometers for different purposes, like meat or candy thermometer. Meat thermometers are not quite reliable, not very precise and it takes the time to measure temperature. The 5 seconds are not important when preparing meat, but in a beer brewing process, the time is everything. You can choose a candy thermometer and we guarantee this will the do trick. Just borrow one from your kitchen, but make sure to ask for a little lady permission.

Long-handled spoon:

A very long-handled spoon that will allow you to reach the bottom of your pot is a necessary part of the equipment. This lovable spoon will keep your hands away from the boiling liquid.

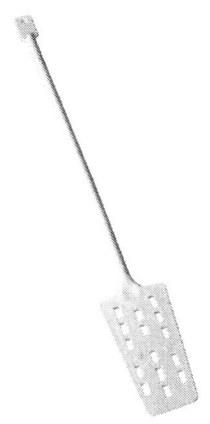

Equipment for fermenting:

Airlock:

At the first glance, the airlock is a funny looking plastic part you may think you do not need. But, do not be fooled my friend, because airlock is a very important part. The airlock is applied on top of a brew bucket or carboy and bubbles away during the fermentation process. Airlock comes in two shapes; S-shaped airlock and three-part airlock. Most brewers prefer three-part airlock for the first fermentation and the S-shaped airlock for the second fermentation. The three-part airlock is loved among beer brewers because it can be cleaned easily. Invest in a good airlock, because this little thingy relieves inner pressure. Without it, some serious beer spill can happen.

Bung:

The bung is used to secure airlock. It is essential if using glass or plastic carboy fermenter. Bung can be made of hard nylon, rubber, or as a wood stopper. To make it shorter, it is a freaking plug, but you need it.

Fermenter:

A fermenter is usually a plastic vessel you use as a safe place for the wort that turns into a beer. As mentioned, you can use the plastic one, but many brewers like to use a glass one. Both kinds will work. 2-gallon plastic bucket with a lid and hole is perfect for a primary fermenter. Just make sure it is a food-grade bucket, and that lid snaps on tightly. Second important is to have a rubber lined drilled hole, a place where you will insert an airlock.

1-gallon jug with stopper and a hole is a second fermenter you need to have, used, naturally, for the second fermentation. Green glass jugs are the best since they protect beer from a sunlight. Make sure the to get a stopper that fits perfectly and the one that has a hole since you need to insert an airlock.

Hydrometer:

Another weird looking, scientific gadget, but important in a beer making process. This little thing shows you an alcohol percent in your finished brew.

13-inch Auto-siphon:

Again, the odd-looking thing you need to transfer the beer from a bucket to a jug and use again to fill bottles with beer.

Equipment for bottling:

Bottling Bucket:

To bottling Bucket, or not to Bottling bucket, that is the question. A bottling bucket is like a plastic fermenter, but with a spigot toward the end. Some like to say you do not need it, but believe us, you do. A bottling bucket is not mandatory, but it is extremely needed for a clearer, and cleaner bottling of your beer. The bottling buckets are cheap and will make your brewing life much easier. The bottling bucket ensures there is no sediment in bottles.

Bottle filler:

Sometimes called a bottling wand, is a hard-plastic tube with a spring-loaded tip used for filling bottles. When the spring-loaded end of the tube is pushed against the bottle bottom, the beer will flow. With the bottle filler, you will be able to fill the bottle without spilling. So that is it, and "thingy" you need to have when filling bottles if your hands are shaking.

Bottle Filler:

Bottles:

You can go crazy with these and maybe choose some unusual bottles, but they need to be made of glass and with some color. The best bottles are dark in color, made from thick glass so let this be your guide when buying beer bottles. The important thing to have in mind is to use plain bottles, and not a twist-off bottles because it is difficult to seal them airtight. Investing in quality beer bottles is a smart thing to do because you can use beer bottles over and over again.

Bottle caps:

Bottle caps are essential, just like the beer bottles. With each filling of the bottles, you will have to use new bottle caps. Luckily, they are cheap.

Bottle capper:

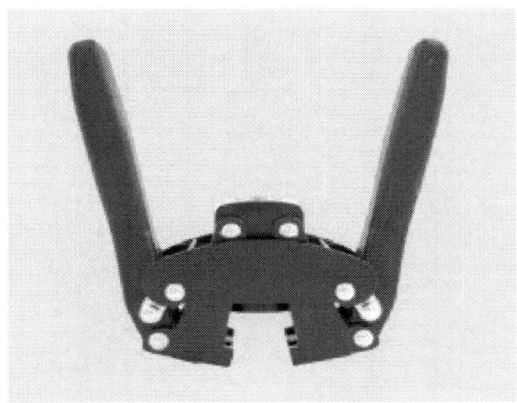

Choose butterfly bottle capper. This is the easiest and cheaper bottle capper. The butterfly capper has a very simple mechanism, just place it on the bottle with a bottle cap, and press down the arms. The mechanism will crimp a bottle cap around the bottle.

Cleaning equipment:

Classic cleaner:

Unscented kitchen cleaner is necessary to clean brewing equipment. Only clean equipment will prevent your dear beer from spoiling. It is important to avoid scented cleaning products (regardless how much you like them) because those scents can stick to the brewing equipment. Instead of perfect beer you can get a florally scented beer, and that is something no one would appreciate.

Sanitizer:

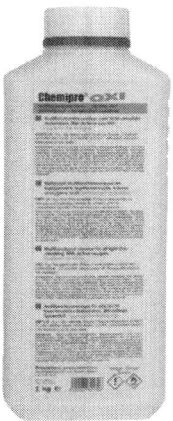

Besides classic cleaner, sanitizer will become your next best friend. Sanitizer ensures there are no microorganisms on your brewing equipment that can also spoil your beer. Some like to use dilute bleach as a sanitizer, but that is not a happy idea. No-rinse sanitizer from homebrewing shops is ideal and very easy to use, with maximum results. The ideal way to use sanitizer is with a spray bottle. This effortless method will do give the best results.

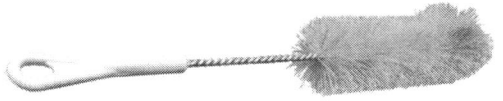

Bottle brush:

Since bottles are hard to clean with plain water, you need to have an efficient method and a tool to clean them. A bottle brush will do a quick job of dislodging any sediment inside bottles that may contaminate beer and make it spoil. Clean and sanitized bottles are important in beer brewing because only clean bottles (with the rest of equipment) will allow you to enjoy your beer.

Additional equipment:

Measuring cup:

Just like with any cooking, beer cooking needs to be precise. Any kind of measuring cup can be used for this purpose. Ideal ones are made of Pyrex glass, but plastic one will also be useful.

Strainer:

When transferring the wort from a boil kettle to the fermenter, some brewers like to strain it through the strainer. This way they avoid hops and similar solids from being transferred. A fine mesh strainer or straining bag are your options.

All about ingredients

You may thing beer has a lot of ingredients, but it is not true. Beer is a simple (outstanding) drink, with a complex aroma, extracted from simple ingredients. In beer making, there are four key elements, malt, hops, yeast, and water. Brewers like to argue which is the most important, but it is important that all four are in even quality. You cannot have a quality beer without quality hops, but what is a quality hop without a quality water? Not so quality beer.

Malt:

The malt is a first and essential ingredient in your beer. There are different types of malt, like corn or rice, but the two most popular types are wheat and barley malt. Malt represents a leftover product of a cereal that has been dried, allowed to sprout, air dried, and heated in an oven. Depending on their expertise, brewers have a choice either using raw grain (and malt it themselves), buy malted grain, or use a malt extract. The easiest to use for the beginners is the malt extract. The malt extract is the extraction of the malted grain and can go directly into the boil. Malt extract does not allow you to add your personal touch, but you will end up with a respected beer. At this point, as a beginner, you may want to impress a friend with a personalized beer, but it is better to have a quality beer made with a malt extract, rather than playing with your luck and raw grains. As stated, the malt extract is ideal for new brewers. Malts can be divided how they are used:

- Base malts
- Specialty malts

Base malts include:

- 2nd row – it is clean and sweet with a mild malty finish. You can use this one for most American Beers.
- Pale malt – has biscuit/nut flavor. A great malt to use in American Beers.
- Pilsen malt – this is the lightest of all malts, with sweet and clean flavor.
- Maris Otter – a malt with a body. It has biscuit/nut like flavor and is great to use for beers with a full body. This is an English style base malt, so it is great to use for English style beers.
- Vienna malt – is rich, aromatic, and deep orange color. It has distinctive warm and malty flavor. Ideal for Vienna beers.
- Munich malt – it has a very robust, malty flavor and is an excellent choice for dark and amber lagers.
- Wheat malt – is used to make wheat and wizen beer types. Wheat malt is used in brewing just long as barley with same diastatic power. Malt wheat is used

for 5-70% in mash, depending on the style. Because wheat has a less outer husk, it has fewer tannins. It is smaller than barley and contributes more protein to the beer, helping in the aid retention.

•Rye malt – lend a dry, spicy flavor with a pale straw color. When making a rye beer, 20% of the malt is used in the mash.

Specialty malts include:

•Crystal malts are made from barley grain, just like base malts, but with a special difference. Crystal malts are steeped and germinated (like the basic malt) but also stewed. This process caramelizes some of the sugar inside. Crystal malts are differently named; Crystal 10L, 40L, 60L…etc., where L stands for Lobovind, a color measure for the malt. The higher number, the darker malt.

•Chocolate malt – the color ranges from 200-500L. Home brewers use this malt for "pimping" the color of their beer. The chocolate malt is made by gently roasting kilned malt to brown it. This step creates a deep, bitter, and roast flavor, compared usually to a chocolate (how obvious). Besides color, it provides a key component in brown ales.

- Black patent malt – also called just black malt is a special type of malt, but not very often used. Some brewers are afraid of this malt due to its rich color and aroma, but just because you are a beginner does not mean you cannot play with it. When properly used, nothing can replace black malt for what it lends to a beer. The black malt primarily gives a high roasted flavor, that carries some bitterness and acidity by the way. Some types can even show a bold fruity character that resembles currants, sultanas, or even blackberries. Most importantly (even in small quantities) it provides a drying quality that brightens up the finish of any beer.
- CaraPils – is specialty malt that was misunderstood for a long time and confused with pale caramel malt. This malt is a dextrin-style malt that constantly increases foam, improves head retention (gives nice foamy head) and enhances mouthfeel, without adding new flavor or color to the beer. This very pale crystal malt leaves unconverted starches that the yeast cannot ferment and this is what gives a greater beer body and foam stability. Can be used alone or with other types of malt. The CaraPils malt upgrades all beer types, including light colored beer.

Low usage rate 1-5% in the mash will give desired results. A truly specialty non-colored and non-flavored malt.

If you still want to try out your luck and use real malted grain, make sure you get milled or crushed grains before purchasing.

NOTE: When to add malt?

To avoid potential mistakes, brewers like to delay the addition of majority of their malt until late in the boiling process. The malt must be added late enough to avoid darkening, but still early enough to be sure the extract is sterilized. Boiling the malt extract 15 minutes is a perfect balance.

Hops:

"Humulus Lupulus"!

No, you are not enchanted with an old spell…this is a Latin name for the spice of beer. Beer made just with malted extract, water, and yeast would be sickly-sweet and dull. The hop is the spice of a beer, that balance flavors and gives that well-known depth.

Who would say you need flowers for a beer brewing? If you did not know, hops are flowers, but unlike those girly flowers, these are real macho flowers. These green buds give a perfect counterbalance to sweet to a sweet, malty brew.

Hops have two very useful things for beer brewing; resins and oil. Resins are those to blame for that distinctive bitter flavor, that marvelous flavor that twists your tongue and excites your taste buds. Different types of hops have different amounts of the resins. Resins are measured in terms of alpha acid percent – AA% and this is what you can see on the boxes of the hops in a beer supply stores. This is ideal for beginners because it tells you bitterness level and that bitter situation you are getting into.

Besides resins, oils are also exciting. The oils can give a full spectrum of aromas, all depending on the exact variety of hop.

Here are some of them that you may find interesting:

Amarillo hops – flowery-citrus like aroma, with medium bittering and 7-11 AA%, used in Ales and IPAs

Cascade hops – flowery, spicy, citrus-like quality, with a slight grapefruit characteristic, with 4.5-7% and used for Porters, IPA, and Pale Ales

Chinook hops – slightly spicy, but very piney hops. 12-14 AA% and is used in most beer styles, from Pale Ales to Lagers

Northern Brewer hop – used mainly for bittering, in combination with the other types of hops. Spicy with a slightly flowery and nutty aroma, mainly used Hefeweizen, IPA and Pale Ale

Willamette hops – AA% 4-6, the King of aroma hops in the U.S. with modest bittering value, with a noble blend of fruit, flower, and earth, and spice notes. Used for Ales and Lagers

In general, hops can be divided into three categories:

- Aroma hops – the ones with lower AA% and an oil profile associated with good and quality aroma. These hops are used as finishing or conditioning hops
- Bitter hops – with stronger AA%. They are used in a boiling process to extract bitterness
- Dual-purpose hops – as the name says they have a dual purpose – they have the qualities of both aroma hops and bitter hops. A perfect example is Northern Brewer hop. Perfect aromatic and bittering hops

Leaf vs. Pellet hops

You probably think "What is with all this hops info, gimme a break", but understanding hops is a MUST in a beer brewing job.

Having fresh hops would be ideal, but these green pine cone shaped flowers are delicate and spoil quickly. Unless you are growing your own, the closest you can get to fresh hops is buying a package of frozen whole-leaf hops.

On the other hand, pellet hops may be the right solution. The pellet hops are compressed hop flowers and are much more widely available. They look weird (like some animal food) but are a great solution because of they last longer, and more affordable than whole-leaf hops. Just because they are in a pellet form, it does not mean they cannot be used to make a quality beer. The procedure of creating a pellet hop does not influence their quality and therefore they will flavor your beer with no excuses. Anyway, most beer recipes call for pellet hops.

NOTE: When to add hops?

Hops are usually added to the beer during the hop boil – a time after the grains are mashed, and the wort is strained, just before adding the yeast. Hops resins are very stubborn and need a good hour-boiling period to properly dissolve and release that unique bitterness. On the other hand, hop oils, need a little time to add aroma and flavor and they would completely be destroyed with a long boil.

See the dilemma here? This is the reason why you need to know your hops!

To solve the dilemma, hops are added in few additions, during the hop boil. Are you surprised? Let us explain; hops added at the beginning are the one to "blame" for lovable bitter aroma – so they are used for bittering the beer. Hops added in the middle, provide both – bitterness and flavors (a great combo of resins and oils), and hops added at the end are all about oils, wonderful flowery, or citrus aromas, and with zero resins that may give any additional bitterness.

To repeat once again; hops resins – bitterness, hop oils – aroma and wonderful flavor.

Yeast:

Besides bread making, yeast is an active member of beer brewing crew. These little organisms are what transform your sugary and malty wort into an awesome beer.

Yeast consumes sugar particles and gives a carbon dioxide and alcohol in return. Within a few days, yeast will eat all the sugar, even the complex sugar compounds, and later slide into a snoozy state of hibernation. This means you have made beer!

Yeast comes in two basic forms;

- Dry and,
- Liquid yeast

Dry and liquid yeast both have their advantages and disadvantages. Selecting a yeast depends on your beer brewing plan and needs.

Dry yeast is sterile, strain-pure and potential to produce great beer. Because it is dried, the shelf life is bigger but is also more tolerant to a warm storage. Dry yeast is packed with nutrient reserves and is ready to use directly without a yeast starter.

The downside of dry yeast is that not all strains can survive the production process, so there are far fewer yeast strains available for beer brewing.

Liquid yeast, on the other hand, has a wide range of available strains. Any strain can be collected and cultured for use by homebrewers. Being a live culture, liquid yeast is far more expensive and is much more perishable. Liquid yeast has a 3 months shelf life and can be destroyed at 90F.

The dilemma is which to use? Well, you can use whichever you want, and there is no reason to pick sides. Use whatever you find easier and affordable.

GOOD TO KNOW: One packet of dry or liquid yeast is enough to ferment 5 gallons of beer.

NOTE: When to add yeast?

Yeast is highly temperature specific. Before you add the yeast, you need to check the temperature is ideal. Achieving the ideal temperature is easy with a cold-water bath. So, before adding (pitching the yeast), the wort must cool to under 80 degrees. Remember – no yeast, no beer!

Sanitation is very important at this step. Once the wort stops boiling, and contact with unsanitized surface can cause contamination, and result in a beer spoil or serious off-flavors in your beer. So, clean and sanitized utensils, yeast and the right temperature will give desired results in a form of great beer.

Water:

Water is very important part of the beer. After all, beer is mostly water and we cannot emphasize how a quality water is a must in a beer brewing. This does not mean you need to use pricey water, as long as your tap water is quality, but if that quality is absent, they at least you have to filter it.

A different water quality and mineral composition will affect your beer. The minerals in water can affect the starch conversion of the mash, but once the sugars have been produced, the effect of water on the flavor of the beer is greatly minimized. So, can you make a good beer with plain water? If using the malt extract, the answer is always Yes, but if brewing with grains, the answer can vary from sometimes to absolutely.

When choosing water, you should check the water ions or the mineral content. The water is not just about using a soft or hard water. Water affects the beer in three ways; it affects the beer pH, it provides "seasoning" from the sulfate-to-chloride ratio, and can cause off-flavors from chlorine contaminants.

Key ions that impact your brew are:

- Calcium - ideal levels are 50mg/L – 150mg/L. Calcium affects the hardness of water and can lower pH during mashing. It also promotes stability and clarity of the final beer.

- Magnesium – ideal levels are 10-30mg/L. The magnesium is needed to aid yeast, but too much can cause bitterness.
- Carbonate and Bicarbonate - different levels for different types of beer. For pale beer, ideal level is 25-50mg/L, while darker prefer 100-300mg/L. These affect alkalinity of the brewing water, and acidity of mash. If levels are too low, the mash will be too acidic, or too high, which makes it at the end inefficient.
- Sodium – in small levels, sodium will have little to no at all impact on the flavor of the beer. Still, sodium contributes to the body and mouthfeel, if in an acceptable range from 10-70mg/L. If over, sodium may cause a metallic taste.

To avoid potential problems, you can use reverse osmosis water. This water allows you to build back in the minerals and create your own desired ion levels.

A Quick Peek

Timeline to brewing beer

Brewing a beer is not an overnight affair, and needs your full attention. As you know with all relationships, this one needs to be special too. From a brew day to opening and enjoying your beer, you need to wait (just like in a real relationship). The usual waiting period is five weeks, with the majority HANDS OFF.

We will prepare you for the beer brewing, but as a start let us give you a peek into the brewing plan, and what can you expect over the next five weeks.

Brew day – it will require 4-5 hours in total (or less if using a malt extract)

Primary fermentation – a 1 week after

Secondary fermentation – 2 weeks after

Time in bottles – minimum is one week, but you can prolong it up to a one year in perfect conditions

TOTAL: five weeks from the brew day to the consummation day

So, simple equation: <u>sugary liquid + yeast + time = perfect beer</u>

Here is a quick and rather dirty and messy overview of what happens during each stage.

The Brew Day:

The beer starts off life as an exciting combo of malt extract or malted grains, water, hops, and yeast. The brew day is a day when you combine all those and transform into a rich liquid, that will in just five weeks turn into a drinkable beer.

The Brew Day is divided into four stages. This is the part where you will do most of the work.

The timeline of a Brew Day:

- Mash – around 1 hour
- Sparge – 1 hour
- Hop boil – 1 hour
- Cool down – 1 hour or more
- Transfer to a fermentation vessel

The first two steps are irrelevant if using a malt extract. As mentioned earlier, the beginners like to use the malt extract because it is an easier method. Just because it is easier it does not have to mean the beer will have a poor quality, it just means you will spend fewer hours making your favorite beverage.

Still, if you want to try a classic brewing method, we will share that secret too.

The first hour; make the mash

To make a mash you need to combine crushed grains with a good amount of water. This is a genuine mash and will look like an oatmeal. Let the mash steep for an hour at the steady temperature. During this procedure, the starches will dissolve in a water and naturally occurring enzymes will break these starches into fermentable sugars.

Second hour; the sparge

The sparge procedure is all about separating sugary liquid also known as the wort. At this process, you need to rinse every, last bit of sugar from the sticky surface of the grains. This process is called sparging. The grains you get are usually used to feed animals. The wort is used later.

NOTE: When using a malted extract these two steps are not to be performed. Instead, you just have to combine malted extract and water to create a wort, and bring that to a boil.

Third hour; the hop boil

This is the easiest and effortless step of the beer brewing. Fill your brew pot with prepared wort on the stove and bring it to a boil over high heat. Once boiling, you can add the hops. The hop boil is usually 1 hour. The hop adding is divided into three parts; at the beginning, another dose midway through, and a dose at the very end. The biggest concern is to track a time, so you can add the hops at the right time. Also, the wort needs to have a full, rolling boil. Only this boil will extract resins from the hops. And that is all you need to do.

Fourth hour; the cooling

As your boiling part is done, you need to cool the wort to a room temperature and get it into the fermentation stage. This helps reduce the risk of bacteria growth and spoiling the freshly brewed beer. The faster way to cool down the wort is with an ice-cold water bath. With this method, you will be able to cool wort within 30 minutes.

4 ½ hours – the transfer

Once the wort is cooled, transfer it into a fermentation bucket. You can strain the wort or siphon it from your boiler. Use a hydrometer to check the gravity and adjust the water or extract. Beer gravity is a total amount of the dissolved solids in water, or in this case the sugar. The gravity reading can give a brewer an idea about final alcohol levels, e.g. imperial stout has original gravity (OG) at 1.080, meaning it has more sugar molecules than ordinary bitter 1.032. This is the reason why the stout has more alcohol.

Once transferred, check the temperature. Once the temperature is below 80F, add the yeast. Froth it with a sanitized whisk and fit the fermenter lid and airlock.

This is just a quick peek of the beer brewing process to inform you what does it take, and how long. In the next chapter, we will explain a fully detailed beer brewing steps.

The Brew Day

Brew, Walk and Talk like a genuine Brewer

And here comes the Brewing day! We have prepared you for this with the previous chapters and there is no need to be scared. Anytime you have doubts, just go back and read the instructions carefully.

As mentioned earlier, beginner brewers like to use the malted extract because it is the easiest way to make a beer. If you want to try brewing beer from a "scratch" with real grains, feel enough confident, or simply want to give it a try, we will describe this method too.

Let us begin:

Lesson one – all about mash (only if using grains)

Brewer's special terms:

- Strike water – a water you mix with grains to get mash
- Strike temperature – the water temperature before adding grains

The mash is a mixture of crushed grains a water. It resembles oatmeal. The best ration is 1lb. grains and 1 ½ quarts water, with the ideal mash temperature of 148-153F.

Mashing is a very important step in a beer brewing and is beyond watery porridge. The process itself has some strict rules, with emphasize on the temperature and the steeping process. It takes around 1 hour for the enzymes to convert starches into sugar.

Action and reaction: what happens?

Yeast: Prepare your yeast. If using a liquid yeast bring it to a room temperature and use as described

Heat source: make sure you have a steady heat source because heating is very important.

Measure the water: make sure you measure your water with a full precision. Use filtered water, or tap water if it is a quality one. Pour water into your boil pot.

Strike water: "strike water" is water heated to 160F or water with perfect "strike temperature". Warm it over a heat source.

Add grains: if using grains, you will follow this step. If using malted extract, this step is not an option. Grains need to be crushed before adding. Once you add your grains, give them a stir and create an oatmeal-like mixture. At this point check the temperature; if it's above 153F, give it a stir to cool down. If it's below 148F, warm the mixture a bit.

Cooking: Place the pot over a steady heat source, so you keep your temperature even. Check the temperature every 15 minutes using a thermometer. Stir the mash to cool it, or leave over heat to warm it. Just remember, 148-153F is ideal. Cook the mash 1 hour. After one hour, the mash should be done. If you had trouble with temperature, you can cook the mash 15 minutes additional, just to make sure all the sugars are extracted from the grains.

Tasting and testing: you can check the mash with two methods. First one relies on your senses. If you feel the mash is very sweet, smells malty, and has a dark-colored look, then your mash is perfect. The second method is a scientific one, and you need to have an iodine tincture. Place one to two drops of the cooled mash into a clean ceramic bowl. Add a drop of iodine. If the iodine color ranges from yellow to amber, the mash is ideal. If it turns purple, cook the mash 15 minutes additionally.

Mash issues

Is mash temperature not steady?

Mash has a thick, porridge-like consistency. This is the reason why mash creates hot-spot because grains settle to the bottom of a pot. To solve this, just stir the mash gently. If warming a mash over a non-steady heat source, warm it for few minutes, then take it off. Stir the mash and check the temperature. Repeat until the mash is ready. Warming mash in intervals will do the trick and prevent heat going up or down too much.

Mash temperature over 153F?

If your mash temperature increases over 153F it is not a big issue if noticed soon. Just stir the mash to release temperature and hotspots. The higher temperature may decrease alcohol content in your beer, but will not spoil it, or make it un-fermentable.

Mash temperature over 170F?

Now, this is the problem. Temperatures this high stop all enzyme activities, meaning you cannot get more sugar from the grains. If this happens in the last minutes of cooking, the beer will be perfectly fine. Just bring down the temperature ASAP. If the temperature is above 170F for 5 minutes or more, you might not have enough sugars to ferment the beer. You can taste it and if it is sweet and has a malty smell, then it is okay, but if you end up with a mixture that tastes like starchy water, well my friend you have messed it up. Better luck next time, and watch over your mash!

Mash temperature below 140F?

This is perfectly fine. Just heat it to hit the 150F ASAP. Again, leaving your mash at this temperature may result in a thinner-bodied beer and higher alcohol levels than expected.

Most important this temperature will not affect your fermentation.

Lesson two – all about sparge (only if using grains)

Brewer's special terms:

- Mash out – raising mash temperature to 170F, to stop enzymatic activities
- Sparge water – fresh water, heat to 170F, used to rinse and residual sugars from the grains, and later added to beer wort
- Sparging – the action of rinsing residual sugars from the grains with "sparge water" or fresh water.

Some include the process of separating wort into sparging process.

•Wort – sweet liquid made during the mash and collected after rinsing grains. Wort is basically an unfermented beer

Action and reaction: what happens?

The mash out: a simple process achieved just by raising mash temperature to 170F in order to stop the enzymatic process. This temperature also loosens up sugars, but it is very important to keep the temperature, not above 170F. Just a one minute will be enough.

Separate wort and grains set a large strainer over your fermentation vessel and pour the grains into the strainer. Allow the wort to drain freely into the fermentation vessel. Be careful at this point, because the wort is very hot. You do not want to sip your beer with a long straw.

There is a special trick that will solve any potential issues; it is called a brewing bag. A simple bag where you place in your grains, and once steeped, just remove the bag. If you prefer a traditional brewing, leave the bag out and do it on a heavier, but more pleasing way.

Sparge: this is an additional step, linked with grains. Pour clear water over the grains to rinse any residual sugar. We need that sugar for a quality fermentation. Once this is done, grains can be discarded or composted. Sparge is important to step is making beer from the scratch as there are many sugars left behind the steeping process (mash process). You have worked really hard to extract those sugars and there is no point leaving them behind, especially because these sugars will feed yeast and make your beer ferment properly.

Proper sparging:

Mark the sides of your fermentation bucket with 1 gallon and 1 ½ gallon measures. This will be a very helpful if your fermentation vessel does not come with the marks.

The process:

Warm the mash to 170F aka. Mashing out. Make sure you gently warm it over medium heat. Stir the mash gently to remove any hot spots. Do not play with the fire and do not increase temperature rapidly. Just do it very easy for the best results.

In a meantime, warm the sparge water. The heated water will remove sugars stuck to the grains. Heat 1-gallon water, but notice the water will vary if making more beer at once.

Separate wort from grains by pouring mash through a strainer. Position the strainer over fermentation bucket.

Sparge the mash by pouring heated water over grains. Recirculate the wort through the grains after. You will do this by quickly cleaning the brew pot and transferring the strainer filled with grains over the cleaned pot. Pour the wort over the grains again. Set the strainer over the bucket once more, and slowly rinse the grains once again with wort. Transfer the strainer with grains over another clean vessel and let any remaining liquid drip into the vessel. You will collect at least of 1 cup. You can add this during the hop boil which is the next step.

Once this process is done, place the pot with the wort over heat and you are ready for hop boil!

These steps can be avoided if using malted extract. Only if you feel confident enough and believe you have what it takes, you can brew beer with grains, otherwise, stick to the malted extract first few times.

Sparge issues:

Mash temperature went over 170F?

This is not recommendable, but not the worst thing that may happen to you. You might notice some extra bitterness in your finished beer, but that is all that can happen.

Was water not sparged?

Try to avoid this, but if it happens, remove the sparge from heat and cover it to keep warm. In a meantime, warm your water to 170F. Continue as described earlier.

No marks added at the vessel?

If you buy a vessel without marks, and you forgot to add marks, then you have to rely on your eyes. When your 2-gallon bucket is ¾ full, you have enough wort.

Recirculating is going very slow?

If this process is taking too long, it is most likely that your wort will cool down too much. This will make it thick and syrupy. To get on a track, rinse the grains with a cup of warm water. If you add the water, extend the hop boil for a few extra minutes. The water will vaporize and you will be on the track again. Also, you can skip recirculating wort, which will leave you with some sediment in your wort, but that is not a huge issue overall.

Lesson three – making wort from a malt extract

Brewer's special terms:

- Wort – sweet, amber liquid extracted from malted barley that the yeast will ferment into beer

Two previous steps can be skipped if using a malt extract. Creating wort with a malt extract is very easy and is done within no time. It does not require grains, mashing and sparging and the malted extract will reduce a risk of all going wrong to a minimum.

Action and reaction: what happens?

Assemble ingredients: gather all your brewing ingredients but at this point only water and malt extract. Also, make sure all your equipment is clean and sanitized.

Heat source: as with the grains, and mashing process, making wort with a malt extract requires a steady heat source.

Making wort: now, this is a fun part where you combine water and malted extract. It begins with heating water and bringing it to a boil. If brewing 3 gallons beer, you need to boil 1-gallon water in a pot. Pour the water into the fermenter and allow it to cool. Bring 2 ½ gallons of water to boil in a brew pot. Stir in 1lb. liquid malt extract and 1lb. dry malt extract, or just use all liquid malt extract. Remove from the heat, and stir malt mixture into fermentation bucket, to make the total 3-gallons.

And that is about it. No need for any additional steps, because the malt extract frees you of grain grinding, mashing, and sparging.

Wort issues with malt extract:

Wort does not ferment?

Worts made from malt extract may sometimes have fermentation problems. The problems occur because of low free amino nitrogen level characteristics of the wort, caused by the poor malt extract. To avoid this, buy quality malt extracts and the ones that have values available from a retailer.

Lesson four – all about hop boil

Brewer's special terms:

•The hop boil or simply known as the boil – the "beer cooking" period, usually 60 minutes, during wort is boiled and hop is added

•Hot break – a moment where wort comes to a boil and breaks through the layer of foam that has collected on surface

- Hop schedule – a different time where you add your hop; at the beginning, through the middle and in the last minutes of cooking
- Flame out – a moment when your turn off the flames or remove the boiling pot from the heat. Some beer recipes call for the last hop adding during the flame out.

Hop boil is the part where you add your hops. This step additional sterilizes the wort, making it safe for the further use. Measure your hops before adding them to a wort. Use a precise kitchen scale, because more, or fewer hops will affect the beer flavor.

During the hop, boil leaves the pot uncovered. There are some compounds that need to evaporate during this process. If left in a beer, these compounds may leave some weird flavors. During the hop boil, you will lose around ½-gallon of liquid, and beer recipes are designed for this, so please, do not add any additional water.

Action and reaction: what happens?

These instructions apply to any batch of the beer. Your specific recipe that you follow will tell you the exact time of the boil, amounts, and types of the hops, and a specific time when to add them.

Bring wort to a boil: place the wort in a boiling pot and bring it to a boil over the steady heat source.

Measure the hops: you can measure hops earlier or while the wort is heating. Do not mix the types of hops.

The Hot Break: as wort comes to a boil, the foam will create on the surface. Watch your pot, so the foam does not go over. The Hot Break happens when the wort comes up through the foam and the foam starts dissolving back into the wort. Once this is achieved, you can start adding hops.

Add hops: as explained earlier, hops are added in three batches. Bittering hops at the very beginning (once the hot break is achieved), flavoring hops after 30-40 minutes, and aroma hops ate the very end. The recipe you are following will show the exact amounts of hops and exact time of addition. Make sure you do not need to stir hops since they will dissolve on their own. During the boil, make sure your wort is at a full, rolling boil all the time.

Finish: Once the hops are boiled long enough (around 1 hour), remove the pot from a heat and begin cooling it.

Hop Boil issues:

Wort boil over?

If your boil pot is small, smaller than 12 quarts, your wort can easily foam up and boil over. Watch your hop during this process and if you think the boil over may happen, reduce the heat. Some suggest adding a few copper pennies into the pot to prevent boil overs. You can also try stirring the wort since this method will release some of the heat accumulated. Anyway, do whatever you think is right, just make sure your wort does not over boil. If the spill happens, you will end up with a little less wort at the end, but it will not influence the beer taste or any similar. You can add some boiled water to top it off.

Hops are not added at recommended time?

Add hops ASAP and do not worry too much. You might notice a less bitterness, but beer will ferment just fine.

Added the wrong measure of hops?

If you add more hops, you will have a stronger bitterness. To solve it, just say you want it like that and call it IPA. If you add fewer hops than the recipe says, do not add hops, just say you wanted it that way. Call it a pale ale or amber, and no one will tell a difference.

Did it boil too long?

Unless all wort is evaporated and your boiling pot is scorched, you can continue with the brewing and fermentation process. Your beer will have more pronounced caramel flavor, more bitter flavor, and less hop aroma.

Different AA% (alpha acid percent) of the hops I have and the one a recipe calls?

If this is no less than a few percent, you can use the hops you have. If the percent is more off, you can recalculate it with an (almost simple formula – who would have told you to need a math).

- AA% recipe hop x grams recipe hop/AA% of new hop = grams new hop

In words;

- multiply the AA% of the recipes hop by the number of grams called for in the recipe
- divide the gained number by the AA% of the hop you have, and you will have the grams you need of your new hop. Turn grams with online calculator into the desired measure.

Lesson five – pitch the yeast

Brewer's special terms:

- Pitch the yeast – adding yeast into the cooled wort

- Original gravity – a gravity reading of wort before fermentation begins.
- Gravity reading – a measure of sugars in the beer. Use hydrometer for this

Before adding yeast, it is important to cool the wort as quickly as possible. For this purpose, use a water bath, to cool wort to a room temperature, or at least 75F.

While the wort is chilling, sanitize the equipment. As soon you add the yeast, the fermentation will begin. For this purpose, you need to prepare your fermentation equipment, including fermentation vessel, appropriate lid, airlock, bung, strainer, hydrometer, measuring cup, and a whisk.

Once all is prepared, check the wort temperature. It needs to be room temperature since that is the temperature yeast really love. Yeast is very sensitive and low temperature will not make it alive, while a too warm area or in this case wort, will kill it and reduce significantly. Action and reaction: what happens?

Cool the wort: fill a large vessel with water and ice cubes. Place in a pot with wort and cool 30 minutes or at least to a 75F. The instead vessel, fill the kitchen sink with water and ice cubes. During the cooling, stir the wort with a long handed spoon, perfectly sanitized.

Sanitation is the key: we cannot emphasize the importance of sanitation and each time, make sure you sanitize your equipment. While the wort is cooling, clean and sanitizes your fermentation equipment.

Wort transfer: once the wort is cooled enough, pour it into a fermentation bucket, through a strainer. You need to use the strainer, to remove any hop residues.

Check the volume: remember those measure we were talking about? Those you had to mark if are not added by the manufacturer? Once the wort is inside the fermentation bucket, check the wort volume. If your wort levels are under 1-gallon (if making smaller batches of beer) pour in some clean water. By the clean water we mean sterile enough and the one that will not introduce any bacteria in your wort. The water will gently minimize the final alcohol percent.

Check the wort with a hydrometer: ant it is a time to use your special equipment. Take some of the cooled wort and place into a clean vessel. Pour the wort into sanitized hydrometer tube. Float sanitized hydrometer in the liquid and watch where the liquid hits hydrometer. The number is usually around 1.050. Have you noticed we are using a sanitized equipment? We are doing so because you will return the tested wort into the fermentation bucket.

Add the yeast: add the desired yeast (dry or liquid) into a wort. If using liquid, prepare it with a starter as described on a packaging. Once the yeast is added, you need to aerate it, or put in simple words – to add some oxygen. You will achieve this by vigorously whisk the wort with a sanitized spoon.

Seal the deal: close the fermentation bucket with a lid. Add the airlock. Before adding the airlock, make sure it is SANITIZED. Fill the airlock to the line with water or vodka. If using a 3-piece airlock, insert the floater and cap, if using a bubbler airlock, place the cap on top. Insert the airlock into the lid, and make sure it fits tightly. Place your fermentation bucket away from the direct sunlight and let the yeast do its magic and make some beer. Within next 24 hours, you will notice some bubbles up through the airlock. Check your beer every day, just to make sure everything looks okay. After 5weeks you will have some outstanding beer.

Pitching yeast issues:

Has wort not cooled down within 30 minutes?

This is not the end of the world. The sooner you cool down the wort, the better, but if the wort is not cooled within suggested time, do not despair. Just make sure you use perfectly sanitized equipment within next steps.

Wort levels are not okay?

If the wort is on the line, just add some sterile water. If the wort is above the line, there is no need to remove any wort. Maybe your hop boil was gentle, or you simply added more water at the beginning. So, what is done, is done. Your beer will have slightly lower alcohol levels, and that is the only problem if you can call it a problem.

Forgot to activate liquid yeast?

Yeats and wort should be the same temperature, so if storing a yeast in a fridge, bring it out to a room temperature. Dry yeast can be used from a fridge, unlike the liquid one. If you forgot to take it from the fridge, cover your wort and allow the liquid yeast to come to a room temperature. Activate is as described on the package.

What is with the airlock?

The airlock can give a headache to some because they do not know how to use it. If you fit into that category, here is a simple solution; if using an s" shaped airlock, you need to fill it with water, sanitizer, or vodka up to a fill line, cap it and insert into bucket lid. If using a 3-piece airlock, fill the main piece with water, vodka, or sanitizer up to a fill line, then drop a smaller cup-like piece inside so it settles over the tube, sticking up inside the airlock. Cap it and insert.

Not sure if sanitized properly?

Sorry, there is no app for this or magic guide. Just follow the steps explained on your sanitizer container and hope for the best. Do the cleaning and sanitation carefully and you will probably do it right.

What happens to your beer now?

Primary fermentation

Brewer's special terms?

- Primary fermentation – the first stage of fermentation, during which the yeast does the bulk of the fermentation
- Primary fermentor – the vessel that holds beer during the first fermentation

The first stage of fermentation, right after the beer is made or should we say brewed is called the first fermentation. At this stage, the yeast is most active and lively. The alcohol stays behind in the beer, while the carbon dioxide bubbles away through the airlock. If you haven't inserted the airlock, you would end up with a beer explosion. What a sight!

At this point, the primary fermentation, yeast eats the sugar and creates different flavors, from the sweet malty flavor to something that tastes like the finished beer. Depending on the type of yeast you used, you can expect different flavors. The stage itself is around one week. As the process is coming to an end, the unused yeast, along with some other solids (from the grains - if using, and hops) will settle to the bottom of your fermentation vessel.

It is very important not to open the lid during this process. Fight that little inner voice and do not peek. Peeking will spoil things. We really mean it because you can bring bacteria into your beer that is sitting around in a perfectly sanitized and bacteria-free area. If you are still stubborn and want to know what is happening let us explain:

First 12-24 hours

You will not see too much activity in the air lock. Yeast is still preparing to eat that sugars.

Next 1-3 days

You will see bubbles popping rapidly in the airlock. It means the fermentation is going nicely, and if you smell some of those bubbles, you will notice it smells like something similar – oh yes, it is a beer smell.

Through the week – the bubbling will gradually slow. It does not mean things have taken a wrong turn, but the fermentation is nearly going to the end. At this point, solids are sticking to the bottom of the vessel, leaving a clear beer above.

After the first or primary fermentation is done, you need to transfer the beer into a clean vessel, for the second fermentation.

Primary fermentation issues:

Primary fermentation did not start at the first 24 hours?

It may be that your house or garage is cold. Try moving the beer to a warmer place and give it again 24 hours. Again, if still no activity, yeast may be the problem (expired date or similar). Grab some fresh yeast with a good date, and add it to beer ASAP.

Forgot to insert airlock?

If your beer has not exploded yet, you can clean your mess. Add water to an airlock and insert it as soon as possible. Continue with the fermentation and see what will happen. Hope for the best.

Not sure about the room temperature?

The ideal temperature for beer brewing is 65-75F. If below 65F the fermentation may not start, and if above 75 all up to 95F you will notice slight changes in your beer flavor. The beer will be okay, but with a fruitier flavor. This is not bad, but also not very welcomed.

The yeast stopped working?

This is called a "stuck fermentation". It does not happen to homebrewers very often, but if you notice there are no bubbles in the airlock on the third or fourth day, the temperature may be the issue. Move your beer to a warmer place.

It has been a week and the beer is still bubbling?

Smell the air coming from the airlock. If it smells right, like the beer, then the fermentation was just slow a bit. Just allow it to finish. If the air smells like something rotten, well my friend it seems your beer has been infected. Sorry, but you cannot use this. Make sure you sanitize better next time.

Transferring beer from the primary to the secondary

Is transferring really necessary?

Brewer's special terms:

•One stage fermentation – an option where the beer is left in the primary fermenter, for the length of fermentation – from pitching the yeast to bottling (option for beers that will be bottled within a few weeks)

- Two stage fermentation – an option where the beer is moved from primary fermenter into smaller secondary fermenter (option for beers that will become long-aged beers)

Transferring beer from the primary to secondary is not a really necessary. If you like to get things done faster, leave your beer in one vessel during the entire fermentation stage up to the bottling part. If anyone asks, this is perfectly legit.

The one stage fermentation has certain benefits, including:

- It is less work, hooray!
- You reduce a risk of exposing beer to the oxygen or introducing bacteria and infection.

The bad sides of this method are:

- The beer cannot become imperial stout. As it lays around too long, the beer will develop stale flavor.

A beer made with this method will have a flavor identical to the two-stage fermentation. We suggest that you try both methods, and get used to the two-stage fermentation because you will be able to develop more complex brews. Still, there is no wrong way here.

If you decide to go with the second fermentation, read the next chapter...

Siphoning beer

Just before you start with your secondary fermentation, you need to transfer the beer to a secondary fermenter.

Brewer's special terms:

•Siphoning – a method of moving beer from one vessel to another, with a use of a siphon.
•Racking – moving beer from one container to another, without introducing a lot of oxygen into the system or transferring a lot of sediment.
•Trub – sediment that collects at the bottom of the fermentation container

Siphoning can be performed with a siphoning tube or a rocking cane.

Action and reaction: what happens?

Siphoning using an auto-siphon:

•Assemble your auto-syphon and sanitize it. Also, sanitize the container that you will be siphoning into.
•Place the primary fermentation vessel on the kitchen surface, and place the secondary fermentation

vessel or bottles below, on the floor and a stable chair. Remove the lid. You will probably notice some scum of floaters, but that is perfectly fine.

•Insert the auto-siphon into the container of beer. Slide the auto-siphon along the wall of the bucket until it hits the bottom and holds it there with hand or clamp so it does not move and stir up the trub.

•Place the open end of the tube inside the empty container. If siphoning into the bottle, make sure it sticks to the bottle throat.

•Use your free hand to pull the auto-siphon partway and pump it once or twice. This should start the beer flow through the siphon. Continue siphoning until the beer is almost completely transferred. You can control the float by applying the hose clamp onto the hose.

•Once you have finished removing the tubing from a container. If you transferred into secondary, insert the jug stopper and the airlock. If you have just bottled, apply the cap.

Siphoning with a rocking cane:

•Sanitize the racking cane. After, hold the open end of the tube under a running faucet. When water flows smoothly out the other end of racking cane, tightly clamp the tube closed with the hose clamp.

- Slide the racking cane into the container of beer. Insert the open end of the tube into the new container.
- Open the hose clamp and let the water trapped inside flow into the new container. This should start the flow of liquid, pulling the beer into the siphon. Since there are only a few tablespoons of water and it will not influence a beer, but if desired you can empty it into a cup.

Siphoning issues:

Can't get the siphon started?

Try separating two containers a little further apart.

Forgot to put a tip on a rocking cane?

No need to worry about it. The siphoning will start anyway. The tip is there to stop transferring thrub.

Transferred sediment?

Even if you transfer a lot of sediment, no need to make a fuss about it. Sediment will eventually settle to the bottom of a new container.

Secondary fermentation

Brewers special terms:

Secondary fermentation – the quiet fermentation where yeast works on complex flavors, solids settle out of beer, and the flavors of beer mellow.

Even the most fermentation is done, with the second fermentation you will give your beer some extra flavors. Yeast is still active at this stage and is breaking some other compounds created during the brewing process. The beer also mellows during these next few weeks. The beer is becoming more balanced without intensive alcohol flavor.

Action and reaction: what happens?

Activity is very low at this stage, so you will not see any air bubbles in an airlock. Maybe a few bubbles, but if many appear you might have a problem, which we will discuss in the issue department.

Once the beer is transferred into a secondary fermentation vessel, you need to make sure everything is on the track in the next few weeks. Check the vessel to make sure the stopped has not gotten jostled and there is water in the airlock.

At this stage, you can add some flavors, like fruit or dry hops. The alcohol that is in the beer will pull those flavors and enrich your beer. The alcohol will also protect beer from any bacteria and make it safe for a further use.

So, leave a secondary fermentation vessel for at least two weeks up to two months. Anytime from two weeks you can start bottling your beer.

Secondary fermentation issues:

Is beer still bubbling?

If you see frequent bubbling, there is a possibility that you have picked an infection. Again, you need to rely on your nose. Smell the air coming from the airlock and if it smells like something rotten your beer is no good.

The stopper fell out of vessel?

If you have noticed this, place the stopper immediately. If this has happened and you have not noticed shortly, there is a chance your beer picked an infection. Apply the stopper and give it some time. If the beer starts to make bubbles in an airlock, that smell rather unpleasant, they your beer cannot be used.

A thin layer of scum?

It can be a sign of a minor infection. As long as the beer has a pleasant aroma, it is okay to bottle. A minor beer infection will not cause any issues.

Once the second fermentation is done, you can bottle your beer.

Bottling

The entire idea of bottling is to fill the bottles with the beer with a minimum fuss and mess. As with any step, make sure equipment you are using is sanitized and clean.

Action and reaction: what happens?

A quick summary; mix the priming sugar and siphon the beer on top of it. This gets the beer off the sediment in the vessel and so there is less risk you will transfer in into a beer. Attach the bottle filler on your siphon and insert it into the bottle. Fill the bottles and cap the bottles.

For this you will need:

- Beer, that is ready to be bottled
- Corn sugar
- Fermentation bucket
- Measuring cup
- Auto-siphon

- Tubing and hose clamp
- Bottle filler
- Bottles and bottle caps
- Bottle capper

Before you start, clean and sanitize your entire equipment.

Prepare the priming sugar: bring water to a boil, add the priming sugar and stir to dissolve. Pour the sugar into fermentation bucket and let it cool to a room temperature.

Mix the beer with priming sugar: Siphon the beer into a fermentation bucket. Avoid stirring.

Use the hydrometer: Siphon a little beer into a measuring cup. Take the hydrometer reading since this will be your final gravity. Use the original gravity to calculate the alcohol in your beer.

Prepare for bottling: Place the beer on the working surface and bottles under. Place the auto-siphon inside the bucket with beer. Pump the auto-siphon to start the beer flowing. Use a hose clamp to tightly close the tube. Fit the bottle filler into the open and end of the tube and insert in back in the bottles so the tip presses against the bottom and release hose clamp.

Time to fill: pump the auto-siphon and fill the bottles until the beer reaches the lip of the bottle. Lift the bottle filler to stop beer flowing, and move to the next free bottle. Continue until bottles are filled.

Time to cap: arrange bottle is front of you. Place the sanitized caps on top of the bottles and position the bottle capper over the bottles. Press down the hands to seal the cap.

The beer is bottled and now you can label it as you wish.

What happens to your beer while in the bottle?

At this stage, your beer is carbonating. Th yeast is eating the last sugar, the one you added during the bottling process.

The carbonation process takes between 7-14 days, depending on the room temperature, the yeast activity, and priming sugar you use.

In addition to carbonated, the beer is clearing and conditioning. A little bit of trub is always present in homebrewing beer, so do not worry about that.

Leaving beer to carbonate and to develop during these two weeks is very important. Beer goes through a bit of shock when bottled and if served too early, they may offer some unpleasant Sulphur flavors. These flavors disappear during the two weeks lay-off. The beer flavor will continue to change, usually for the better, over the next few months. You will be able to notice that "older" beer has a different flavor than the one first one you have tried.

With most beer styles, beer flavor improves, so you have a very good reason to leave the beer in bottles.

While waiting, you can try some other types of beer, just to tingle your imagination how your beer will turn out.

Beer Recipes

Classic Lager

Yield: 5.00 Gallons

Ingredients:

- 7 lb. White Sorghum Malt (mash/sparge)
- 3 lb. Flaked Rice, mash
- 1 lb. Flaked Corn, mash
- 1 oz. Tettnanger Whole Hops - boil 90 min
- 1 tsp Irish Moss - boil 15 min
- 1 lb. Light Brown Sugar - boil 1 min
- 1.5 lb. Honey - boil 1 min
- Dry Lager Yeast

Directions:

1. Mash grains at 150F degrees, for 1-3 hours or until starch conversion is over.
2. Bring wort to boil and achieve hot break.
3. Add hops, remaining fermentable sugars according to schedule.
4. Chill wort to below 80 degrees and add in yeast.
5. Transfer beer to plastic bucket fermenter and follow primary fermentation.
6. After first fermentation, it is time for second fermentation. In this process, there is no fermentation, but clearing and conditioning.
7. Beer is set to be packed.

Robust beer

Yield: 3.00 Gallons

Ingredients:

- 6.5 pounds liquid light malt extract
- 1 pound liquid Munich malt extract
- 1 pound Crystal 40L malt, crushed
- 3/4 pound Chocolate malt, crushed
- 1/2 pound Black patent malt, crushed
- 1 ounce Cascade hops—boil 60 minutes
- 1 ounce Cascade hops—boil 15 minutes
- 6 gallons of tap water, split
- 1 Liter starter of liquid American Ale yeast

Directions:

- Place crystal malt, black patent and chocolate malt in the large mesh-grain bag.
- Place the bag in 3 gallons of water, in 5-gallon pot and immerse the grain.
- Bring to heat and try to set the bag so it is not directly on the bottom.
- Remove the bag when the temperature reaches 170degrees.

- Bring prepared wort to boiling; while water is still heating stir in liquid malt extracts and continue stirring until dissolved.
- When the boil begins ad in cascade hops in mash bag.
- After 45 minutes, add remaining 1 oz. cascade hops.
- The mixture should boil for 60 minutes; remove after 60 minutes from heat.
- Cool wort by placing the pot in an ice bath; transfer to sanitized fermenter and add enough cold water to top 5 gallons.
- Carefully pour yeast into cooled wart, somewhere around 70 degrees, and agitate vigorously.
- Cover with an airlock.
- Perform first and second fermentation.

NOTE: Ferment between 65-68 degrees. After 3 weeks bottle.

American IPA

Yield: 7.5 Gallons

Ingredients:

- 9 pounds Light liquid malt extract
- 0.75 pounds Crystal 20L malt, crushed
- 1 ounce Magnum hops - 60 minutes
- 1 ounce Simcoe hops - 15 minutes
- 1-ounce Sorachi Ace hops - 15 minutes
- 1 ounce Simcoe hops - 0 minutes (flame out)
- 1-ounce Sorchi Ace hops - 0 minutes (flame out)
- 11.5-gram package Safale US-05

- 1 ounce Simcoe hops - for dry hopping in Secondary
- 1-ounce Sorachi Ace hops - for dry hopping in Secondary

Directions:

- Tie the crystal malt in a mesh bag.
- Set the bag in 3 gallons water in a 7.5 gallons pot; immerse the grain.
- Begin to heat and when the temperature reaches 170F degrees, remove grain bag from the water.
- Add 3.5 gallons water and bring wort to boil.
- As water is heating, add in light liquid malt extract.
- Stir well until malt is dissolved.
- When the boil begins, add Magnum hops in mesh bag.
- After 60 minutes boiling, add 1-ounce Sorachi Ace and 1 ounce Simcoe hops in a mesh bag, cover and remove from heat.
- Place the pot with wort in ice bath and chill until wort reaches 70 degrees.
- Transfer to fermentation bucket and top off to make 5 gallons using refrigerated water.

•*Use a sanitized auto-siphon racking cane to remove enough wort to take a gravity reading with your hydrometer. Make a note of this number, since you will be using it to calculate the actual alcohol content when it's done fermenting. The reading should be around 1.067.*

•*Pour yeast into cooled wart and agitate vigorously.*

•*Cover fermenter with sanitized air lock.*

•*After 2 to 3 weeks when primary fermentation is complete, transfer to a secondary carboy for conditioning, add 1-ounce Sorachi Ace and 1 ounce Simcoe hops for dry hopping and store as cool as possible.*

NOTE: *Bottle after another one to two weeks using enough priming sugar for a medium level of carbonation.*

Belgian beer

Yield: 5 gallons

Ingredients:

- *9 pounds Pilsner malt extract*
- *1 pound light Belgian candy sugar*
- *1 pound Carapils malt, crushed*
- *2 ounces Hallertau hops - 60 minutes*
- *6 gallons of tap water, split*
- *2 Liter starter of liquid Belgian Ale yeast (Whitelabs WLP500 or Wyeast 1214)*
- *Directions:*

- *Place 3 gallons water in the refrigerator to chill.*
- *Place the caramels malt in large mesh hop bag and tie.*
- *Place bag in 3 gallons water n 5-gallon pot.*
- *Begin to heat and when the temperature reaches 170degrees remove the hop bag.*
- *Bring wort to boil; as water is heating add 2 pounds Pilsner malt extract.*
- *Stir well until malt is completely dissolved.*
- *When the boil begins, stir in Hallertau hops, all 2 pounds.*
- *Boil for 45 minutes and add remaining 7 pounds Pilsner malt extract and 1 pound candy sugar.*
- *After total 60 minutes boiling, remove from heat.*
- *Cool wort in ice bath until below 85degrees.*
- *Transfer to fermentation bucket and top off to make 5 gallons, using refrigerated water.*
- *Use a sanitized auto-siphon racking cane to remove enough wort to take a gravity reading with your hydrometer. Make a note of this number, since you will be using it to calculate the actual alcohol content when it's done fermenting. The reading should be around 1.075.*
- *Pour yeast in cooled wort and agitate vigorously.*

- *Cover with clean lid and ferment in dark place between 68-70F degrees.*

NOTE: Bottle after six weeks using enough priming sugar for a high level of carbonation.

Red ale

Yield: 5 gallons

Ingredients:

- *6 gallons of tap water, split*
- *6 pounds Light liquid malt extract*

- *1 pound CaraRed malt, crushed*
- *1/2 pound Crystal 60L malt, crushed*
- *2 ounces Black Roasted Barley malt, crushed*
- *1 ounce Centennial Hops—60 minutes*
- *1 ounce Centennial Hops—15 minutes*
- *1 ounce Amarillo Hops—5 minutes*
- *1 Liter starter of American Ale yeast (White Labs WLP001 or Wyeast 1056)*
- *1 ounce Amarillo Hops—for dry hopping in secondary*
- *priming sugar for bottling*

Directions:

- *Refrigerate 3 gallons of the water n clean container.*
- *Tie crystal, CaraRed malt and Black roasted barley in a large hop bag.*
- *Set the bag in 3 gallons water in 6-gallon pot; immerse the grain.*

- *Begin to heat, making sure that bag is not touching bottom.*

- *Remove the hop bag when the temperature reaches 170F degrees.*

- *Bring mix to a vigorous boil and as water is a heating stir in 6lb. of malt extract, stirring until dissolved.*

- *When begins to boil, add 1 oz. centennial hop to hop bag.*

- *After 45 minutes of boiling add remaining 1 oz. centennial hop to hop bag.*

- *After 55 minutes is passed, add Amarillo hops to hop bag.*

- *After 60 minutes of boil, remove from heat.*

- *Cool wort to below 85F degrees and transfer to clean fermentation bucket.*

- *Top off to make 5 gallons, adding refrigerated water.*

- *Use a sanitized auto-siphon racking cane to remove enough wort to take a gravity reading with your hydrometer. Make a note of this number, since you will be using it to calculate the actual alcohol content when it's done fermenting. The reading should be around 1.050.*

- *When wort is below 70F degrees, stir in yeast.*
- *Cover with clean lid and ferment in dark room, between 65-68F degrees.*

NOTE: *After 2 to 3 weeks when primary fermentation is complete (take at least two consistent gravity readings), transfer to a secondary carboy for conditioning, add 1 ounce Amarillo hops for dry hopping and store as cool as possible.*

Bottle after another one to two weeks using enough priming sugar for a medium level of carbonation, 0.9 lb. per gallon.

English ale

Yield: 7 gallons

Ingredients:

- *4.75 pounds Marris Otter malt, crushed*
- *0.6 pounds Crystal 40L malt, crushed*

- *0.3 pounds Crystal 120L malt, crushed*

- *0.15 pounds chocolate malt, crushed*

- *0.5 ounces Northern Brewer hops - 60 minutes*

- *6.5 gallons tap water*

- *1 package liquid English Ale Yeast (White labs WLP002 or Wyeast 1968)*

Directions:

- *Line 7.5 gallon kettle with mash bag.*

- *Add 2.5 gallons water and heat until 164F degrees. Remove from heat.*

- *Mash in the 5.8lb grain into the water and stir for 2 minutes; temperature should be around 154F degrees.*

- *Cover the mash and only uncover each 20 minutes to stir; heat 3 gallons water to 185F degrees.*

- *After 60 minutes, mash-out by carefully pouring the heated 3 gallons water into the mash, stirring to equalize temperature to 170F degrees.*

- *Slowly remove the grain bag out of the liquid, allowing the wort to drain from the grain.*

- *Hold the grain bag above the kettle for about 5 minutes as the wort drains. Top the wort off with enough water to 6 gallons.*

- *Bring wort to a boil. When the boil begins, add 0.5 oz. Northern Brewer hops in a mesh bag.*

- *After 60 minutes of boil, remove from heat.*

- *Cool wort by placing the pot in an ice bath and chill to 70F degrees.*

- *Transfer to clean bucket.*

- *Use a sanitized auto-siphon racking cane to remove enough wort to take a gravity reading with your hydrometer. Make a note of this number, since you will be using it to calculate the actual alcohol content when it's done fermenting. The reading should be around 1.032.*

- *When wort is 70F degrees or below stir in yeast and agitate vigorously.*

- *Cover bucket with clean lid and ferment in dark place, keeping the ambient temperature consistent, preferably between 65 and 68°F.*

NOTE: Bottle after 1 to 2 weeks when fermentation is complete, using enough priming sugar for a medium level of carbonation, adding 0.9lb. Per gallon.

California lager

Yield: 7 gallons

Ingredients:

- *5 pounds American 2-row malt, crushed*
- *1.25 pounds Munich malt, crushed*
- *1 pound Crystal 40L malt, crushed*
- *2 ounces Chocolate malt, crushed*
- *1 pound dry malt extract*
- *0.75 ounces Northern Brewer hops - 60 minutes*
- *0.5 ounces Northern Brewer hops - 30 minutes*
- *0.75 ounces Northern Brewer hops - 10 minutes*

- *1.5 Liter Starter of White Labs WLP 810 or Wyeast 2112*

Directions:

- *Line the 7.5-gallon kettle with the mesh bag and fill with 2.5 gallons of tap water and bring to 162°F. Remove from heat.*

- *Mash-in by slowly adding the 2-row, Munich, Crystal 40L and Chocolate malt into the water and inside the bag.*

- *Stir for 2 minutes to creating a consistent mash. The temperature should equalize to about 152°F.*

- *Cover the mash, only uncovering to briefly stir every 20 minutes. In a separate pot, heat 3 more gallons of water to 190°F.*

- *After 60 minutes, mash-out by carefully pouring at 190F heated water into the mash, stirring to equalize temperature to about 170F degrees.*

- *Slowly raise the grain bag out of the liquid, allowing the wort to drain from the grain. Hold the grain bag above the kettle for 5 to 10 minutes as the wort drains. Add the 1 pound of dry malt extract and top the wort off with water to 6.5 gallons.*

- *Bring wort to a vigorous boil. When the boil begins, add 0.75 ounces Northern Brewer hops in a mesh bag.*

- *After boiling for 30 minutes, add 0.5 ounces Northern Brewer hops in a mesh bag.*

- *After boiling for 50 minutes, add 0.75 ounces Northern Brewer hops in a mesh bag.*

- *After a total of 60 minutes of boil, remove from heat. Warning: After wort cools below 180°F everything that touches it should be sanitary, and exposure to open air should be limited as much as possible.*

- *Cool wort by placing the pot in an ice bath or by using a wort chiller until it is at 60°F. Transfer to sanitized fermentation bucket.*

- *Use a sanitized auto-siphon racking cane to remove enough wort to take a gravity reading with your hydrometer. Make a note of this number, since you will be using it to calculate the actual alcohol content when it's done fermenting. The reading should be around 1.047. Cover fermenter with a sanitized stopper and airlock.*

- *Agitate vigorously for at least 5 minutes or aerate using pure oxygen for 1 minute. Add in a 1.5L starter of California Lager yeast.*

NOTE: Ferment for at least 10 days at 60°F. Condition by allowing the beer to rest for 3 weeks at 50F degrees. Bottle after conditioning is complete, using enough priming sugar for a medium to high level of carbonation, 1.5 lb. per gallon.

What to serve with your Beer?

Hot Chicken wings

Serves: 4

Preparation time: 10 minutes

Cooking time: 25 minutes

Ingredients:

- *2 lb. chicken wings washed and drained*
- *¼ cup hot chili sauce*
- *½ cup raw honey*
- *½ cup soy sauce*

- *1 teaspoon garlic powder*
- *1 tablespoon lime juice*
- *4 tablespoons unsalted grass-fed butter*
- *Fresh ground salt and pepper*

Directions:

- *Preheat oven to 420F and place wire rack onto a baking tray.*
- *Season the chicken wings with the salt and pepper and set on wire rack.*
- *Bake for 20-25 minutes.*
- *Meanwhile, prepare the hot sauce; melt butter in sauce pan.*
- *Add garlic powder, hot sauce, soy sauce and lime juice. Season with salt and pepper and bring to boil.*
- *Remove from the heat and pour into large bowl.*
- *Add in baked chicken wings and set aside for 10 minutes.*
- *Serve after with some hot sauce and a squeeze of fresh lime juice.*

Scotch Eggs

Serves: 4

Preparation time: 15 minutes

Cooking time: 8 minutes

Ingredients:

- *4 hard-boiled eggs*
- *¾ pound sausage meat*
- *¼ teaspoon ground cumin*
- *¼ teaspoon cayenne pepper - optional*
- *Oil – to fry*

Preparation method:

1. To hard boil eggs; place the eggs in a pot of water, so they are covered.

2. Heat the water until starts to boil. Boil the eggs for 2 minutes and set aside in hot water for 10 minutes.

3. Transfer to ice cold water and set for 10 minutes more.

4. Peel the eggs, and set aside.

5. Squeeze the sausage meat from the sausage coat in a bowl. Add cumin and cayenne pepper – for the spicy option.

6. Combine with hands and divide into 4 equal portions.

7. Form the kind of plat pancakes from the meat and place the egg in the center. Wrap the egg with the meat, so it is completely hidden.

8. Heat oil in non-stick skillet and when hot enough place in the wrapped eggs.

9. Cook each egg for 8 minutes, until cooked from all sides.

10. Serve immediately.

Spiced Pork belly

Serves: 4

Preparation time: 10 minutes + inactive time

Cooking time: 70 minutes

Ingredients:

- *1 ½ lb. pork belly*
- *2 tablespoons apple cider vinegar*
- *2 tablespoons soy sauce*

- *3 tablespoons raw honey*
- *1 teaspoon Chinese 5-spice mix*
- *Fresh ground salt and pepper*

Preparation method:

- *Place pork bellies in a large bowl and add remaining ingredients.*
- *Cover with plastic foil and refrigerate for 4 hours, in the refrigerator.*
- *Preheat oven to 350F and place the pork bellies in baking dish. Pour over the marinade and some olive oil.*
- *Bake for 60-70 minutes, flipping half way through. You can drizzle with 1 tablespoon honey additionally.*
- *Set on wire rack to cool before slicing.*

Grilled Lamb chops

Serves: 4

Preparation time: 10 minutes

Cooking time: 10minutes

Ingredients:

- *4 lamb chops, around 1 ½-inch thick*
- *2 tablespoons Dijon mustard*
- *½ tablespoon grass-fed butter*
- *3 tablespoons pesto sauce*
- *2 tablespoons olive oil*
- *2 tablespoon finely chopped rosemary*
- *Fresh ground salt and pepper*

Preparation method:

1. Drizzle the one side of pork chops with the olive oil and sprinkle with chopped rosemary.

2. Heat the olive oil and butter in large skillet.

3. Add pork chops on a grill pan, leaving some space between and season with salt and pepper.

4. Cook for 4-5 minutes, until browned.

5. Turn on the other side and cook for 5 minutes more.

6. Spread mustard onto lamb chops and top with pesto.

7. Cook chops briefly for 2 minutes more. Remove from the heat.

8. Flip on the other side and let it sit for 5 minutes.

9. Serve while still hot and pour over cooking juices.

Printed in Great Britain
by Amazon